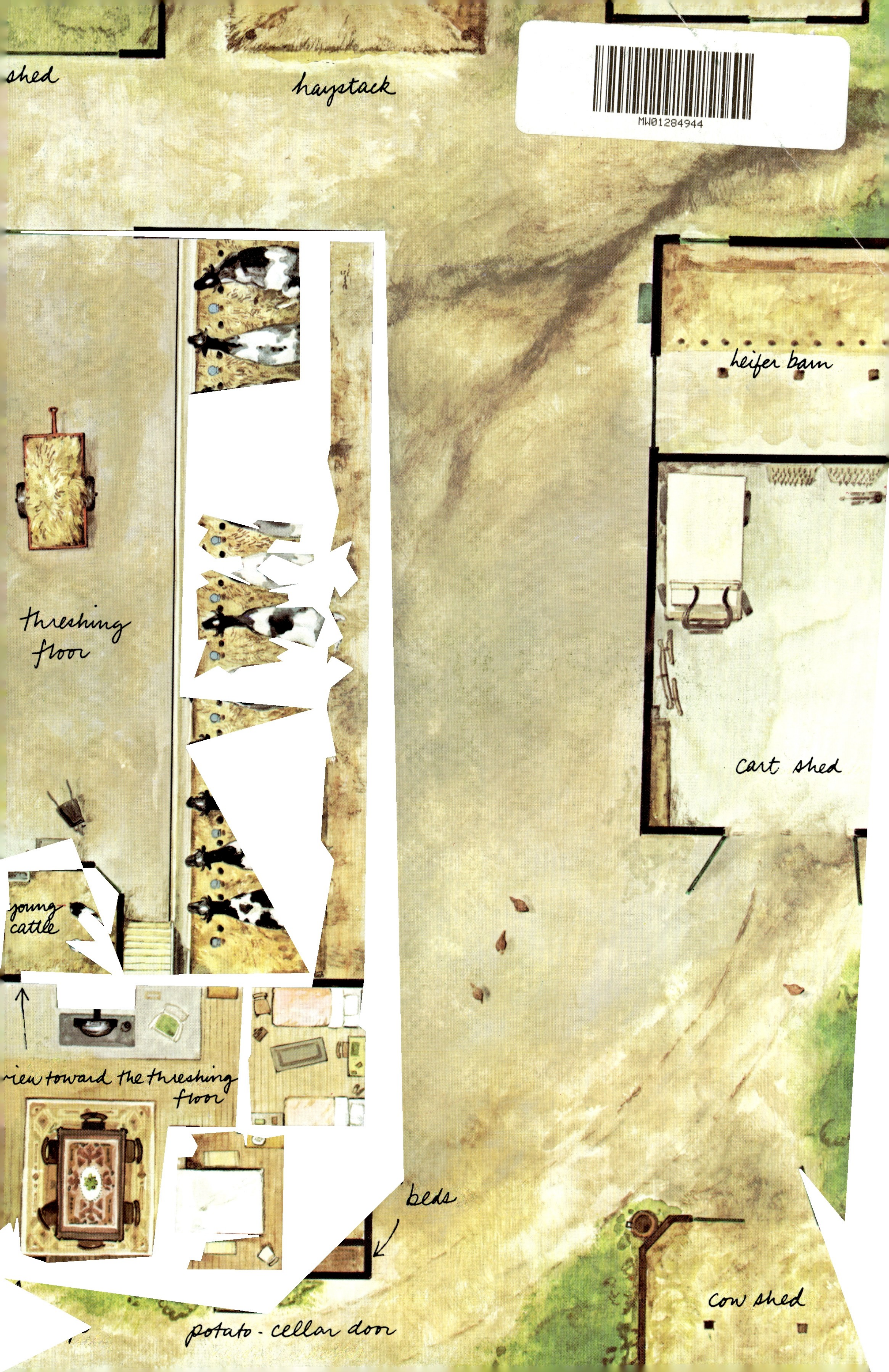
shed
haystack
MW01284944
heifer barn
threshing floor
cart shed
young cattle
view toward the threshing floor
beds
potato-cellar door
cow shed

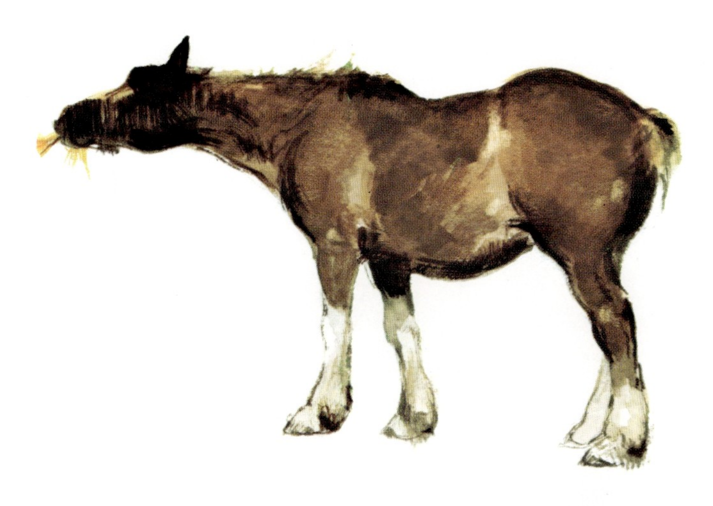

The Farm Book

Introduction by Robert Elman

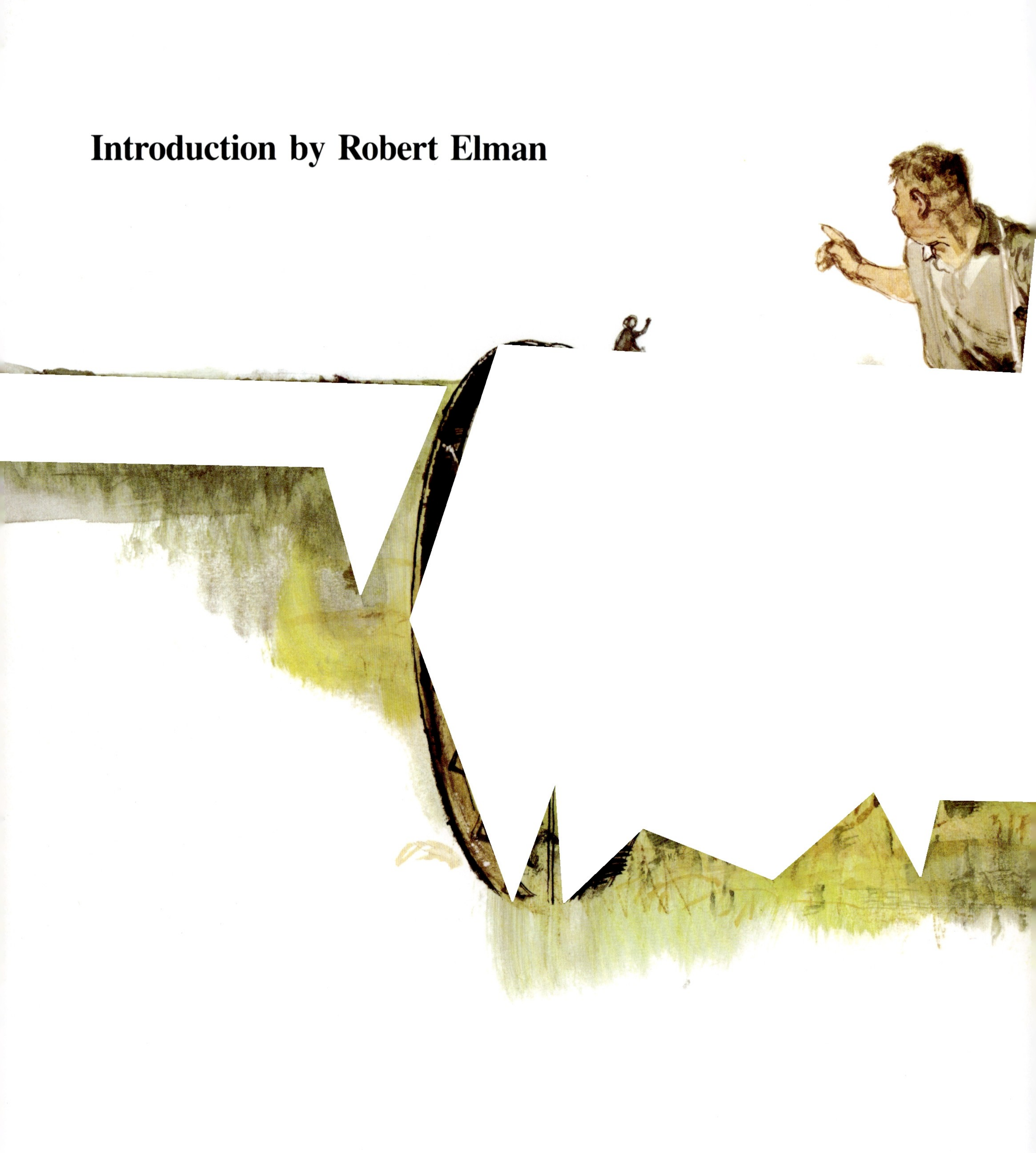

The Farm Book

Rien Poortvliet

Harry N. Abrams, Inc., Publishers

Reprint 1994

Editor: Joan E. Fisher

Library of Congress Cataloging in Publication Data

Poortvliet, Rien.
The farm book.

Translation of Hooi en te gras.
1. Agriculture—Netherlands—Pictorial works.
2. Farm life—Netherlands—Pictorial works. I. Title.
S469.N4P6613 630′.9492 80-14217
ISBN 0-8109-0817-4

Originally published under the title *Te hooi en te gras*

Published in 1980 by Harry N. Abrams, Incorporated, New York
A Times Mirror Company

Printed and bound in Hong Kong

Introduction

Rien Poortvliet is a vivid chronicler of Dutch life and landscape, a genre painter who seldom wanders far in search of subjects for his pen and brushes. Yet his work has been welcomed as enthusiastically in America as it has been in Europe. Clearly the reason is that his seemingly provincial, rural themes are astonishingly cosmopolitan, his insights and appeal international. He is a latter-day Boswell (and occasionally a Hogarth) on tour in his own land, the kind of painter who understands how William Blake could see the world in a grain of sand. The unique has universal implications. More can be seen in a Dutch barnyard than merely one corner of Dirksland or Frisia.

The small Dutch farm is not an isolated phenomenon but a microcosm of a way of life that has existed in one form or another across the globe. It is a way of life worthy of examination today, one that has become rarer and rarer but is being revived—or preserved—by people who feel a personal affinity for the land.

A simple, back-to-the-earth life-style has become a desirable alternative for many urban dwellers who are disenchanted with the city's unnatural crowds, debris, pollution, concrete landscape, purposeless tension, goalless rush, plastic foods, and plastic standards. Those who

are content to live in an urban environment sometimes question whether a return to rural life is not a pathetic retreat from society's obligations—an escape into romantic nostalgia. But it is not a running away; it is a running to. By returning to the land, many people regain control of their lives and find a new pride of accomplishment. These people discover with delight that there is real value—esthetic, spiritual, social, and healthful—not only in the leisure to be friendly but in such simple material blessings as newly churned butter, fresh cream for one's coffee, properly pressed cheese, and home-baked bread.

There are other kinds of universality in Poortvliet's scenes of Dutch farm life. Americans who have seen small farms in southeast Pennsylvania, central New York, or parts of Ohio may feel a sensuous pang of recognition as they look at some of Poortvliet's farm buildings, harnesses, work clothes, implements, tools—even some aspects of the very way of life. Where the similarities are not precisely Dutch they are likely to be at least Flemish; there is, after all, a relationship between the pungency of Limburger cheese and that of New York's Liederkranz.

The small Dutch farm itself, the brush-edged, patchwork farm whose tangled windbreaks and hedgerows are a haven for rabbits and upland birds, has strong similarities to the small farms that once quilted Georgia, Oklahoma, and the Dakotas. Our rabbits may inherit this earth. The quail population of the South and the pheasant population of the Midwest have been diminished by the trend toward "clean farming" and vast, unbroken "single-cash-crop" fields. In Holland, as well, the pheasants suffer where modern agribusiness impinges on old-fashioned farming practices.

One wonders if the people do not suffer, too. Poortvliet would grasp instantly the significance of the advice proffered by naturalists like Hal Borland, who have concluded that the kindest thing to do with the land is as little as possible. In essence, this is a land ethic voiced by many, from Henry David Thoreau to Aldo Leopold. The small-land farmer understands intuitively that nature can be harvested without being ravaged.

Like the farmer himself, Poortvliet is a man of the land. His rapport with animals both wild and domestic is renowned, and in this book he reveals a similar rapport with the robust people who work the land. He is no maudlin outsider looking in. As a boy, he became familiar with the sounds and smells of the barnyard and the stalls, the hayricks and the fields and the kitchen. He knows the hot exhalation of fermenting silage and the steaming of animals as well as the warmth of that kitchen. He also knows the farm where the watercloset is cold and the outhouse colder. And he has no aversion to depicting a farmer in pastures far enough from either watercloset or outhouse to welcome the convenience of a screening hedge while he tends to a call of nature.

If some of Poortvliet's brooding landscapes display a knowledge of the romantic tradition, they are very real nonetheless, and he is a hearty realist in his sketched or painted narratives. We share his fascination as a newborn piglet is cupped in a farmer's hand and dried with straw. We see the sow farrowing—a happy occasion, a normal birth. We are also reminded that a sow may inadvertently suffocate a piglet, and that too large a litter will doom one or two of the young. We see a Cesarian section performed on a cow, and rejoice at its success. We watch a veterinarian using his stethoscope and rectal thermometer, and we observe with wonder how an animal as ponderous as a horse can be managed during surgery for a hernia.

Through Poortvliet's eyes, we see not only the coupling of animals but the courting of people and more than one tryst in a hayloft. No aspect of farm life escapes his scrutiny, and some aspects prompt a sly commentary. An interpreter as well as an observer, Poortvliet follows the humanistic tradition of the Flemish master Pieter Bruegel, who reveled in bold images of peasant life, presenting its joys and pains, viewing man compassionately even as he wittily exposed mankind's follies.

The comparison is not invalidated because Breugel was essentially a pessimist about the human condition and Poortvliet is not. The two painters, separated by more than four centuries, share certain inclinations apart from humor, compassion, and a fond empathy for people of the land. The Flemish and the Dutch seem always to have had a predisposition toward humanism (witness Bruegel and the Dutch humanist Erasmus, among others) and for a melding of the real and the fantastic. Bruegel was an admirer of Jeroen Van Aeken, a Dutch allegorical painter more familiar to us as Hieronymus Bosch. Bosch as well as Bruegel would have taken enormous delight in Rien Poortvliet's illustrations for the famous book *Gnomes,* a wondrous combination of "natural history" and "anthropology" in which the supernatural little creatures provide a vehicle for reflections on man's nobility and foibles.

Poortvliet is, of course, a modern watercolorist, and the intention here is not to liken his technique or accomplishment to that of Bruegel. All the same, there is an inescapable relationship of viewpoint and subject matter—people, animals, feasting, carousing, mating, religious worship, labor, the harvest, and so on. A Poortvliet farm dog does not simply pause beside a milk can but lifts its leg. Bruegel would have smiled and nodded. He would have lauded Poortvliet's use of unexpected angles—the glance upward at the exposed leg of a farm woman on a ladder or a downward view showing the entire layout of a farmhouse, pens, barns, and other outbuildings. And he would have appreciated the use of dramatic color accents: a rusty red fox steals away from a farmer's chicken coops with a red-combed white bird, depicted against a lean, wintry background.

Further comparisons are both inevitable and revealing. Poortvliet, like Norman Rockwell, has successfully carried into the twentieth century the European and American genre idiom of the nineteenth century—which, in turn, can be traced far back to the narratives of Bruegel and even beyond. One notes with interest (and amusement) the critical change in attitude toward Norman Rockwell. Once he was considered to be a mere provider of *Saturday Evening Post* covers and other "commercial trivia." Now he is accorded gallery and museum status. Rockwell is considered consummately American, yet Poortvliet's humorous glimpses into private lives are Rockwellian (which perhaps is synonymous with universal): a grinning laborer reveals his need of dental work; a man naps after a heavy lunch while his dog contentedly emulates him; a boy sits in a watercloset as his faithful dog waits just outside; a farmer balances a heavy milk can on a spindly bicycle; a calf quizzically inspects a rabbit; a portly man squirms in an undersized tub; in church, a cupped hand hides a pocket watch as the pastor drones on, and a man dozes through the sermon while a woman scolds her fidgeting son and another lady tucks up her Sunday coat to prevent it from wrinkling as she sits down; a farmer chases a suspicious chicken; a young couple in a hayloft reveal their panic at imminent discovery; and St. Nicholas arrives with his diminutive helper at a small, isolated farmhouse.

There are still other similarities between this Dutch painter and various American artists. Poortvliet's meticulous attention to the details of farm architecture, tools, and implements is much like Eric Sloane's, though perhaps warmer. His feel for the shape, function, and texture of these artifacts brings to mind Sloane's book *A Reverence for Wood*.

As a landscapist, Poortvliet can be compared to many watercolor virtuosi, including the American painter Ogden Pleissner. His use of composition accentuates his narrative dynamics, and he combines accuracy of form with modern impressionism. Few watercolorists can surpass his suggestion of mist, rain, snow, luminosity, and shadow, the deft translation of nature's subtleties. It may or may not be significant that Pleissner lives in a Vermont farmhouse. It may or may not be significant that both painters enjoy hunting and fishing and the portrayal of these activities—and of the wildlife involved therein. What is important is their sensitivity to man's interaction with nature.

Far too skilled to confuse solemnity with seriousness of purpose (though his work on suitable occasions is stark), Poortvliet draws much of his humor from the outlook of his subjects. The backbreaking nature of farm work demands comic relief.

It is a wry rural humor, neither Dutch nor American but reminiscent of frontier existence in many places and times. Poortvliet comes upon an inordinately long farmhouse on the road from Ziewent to Ruurlo. Before painting it, he inquires within. He discovers the house has been named "Psalm 119" because of its great length. Only the architect is Dutch; the peasant's amusement with life is universal.

Robert Elman

Uncle Dirk and Old Gray

I saw him as soon as I got off the train at the small wayside station: Uncle Dirk!

It was during the war. I was about eleven years old and was being sent to stay with him at the farm. And even though I had never met him I had heard things here and there not intended for the ears of little ones. I sensed that my Uncle Dirk was not exactly a model character.

During the trip from my home in Schiedam to Dirksland I felt uneasy about meeting Uncle Dirk, and I tried to recall everything I had ever heard about him. Some of it made no sense to me but one thing was clear: he did not always live by the word of God; indeed, he might even be a nonbeliever, for he was a freethinker. And there was something else—I didn't understand exactly what it was but it had something to do with one or the other pretty local girl. And, furthermore, he wasn't a very diligent farmer but more of a happy-go-lucky type who liked a practical joke. Yet among other things he was also a member of the village council, the conductor of a socialist choral society, and a member of the local theatrical troupe.

And now I was to meet him. He greeted me warmly, saying "Well, well, so you're Rien!" After all my anticipation, I thought he sounded so friendly that I almost laughed—he made me feel welcome. In an instant I didn't care what I had heard about him. He was cheerful, singing as we walked along, and as I noticed later he sang even while he worked. He gave me self-confidence, and I enjoyed every minute I spent with him. It was a time of freedom and wonderful discovery.

I loved everything—the hearty meals with bacon drippings, the fragrance of apples drying in my sleeping loft, the clop-clop of the children's clogs on the way to school, where we still wrote on slates the local dialect that I rapidly acquired. Saturday evenings, dressed in my best outfit, I was allowed to go for a stroll with Uncle Dirk and his friends or to the choral group, for which he would cut himself a baton from a tree along the way. It was a glorious time, and I learned to love the kind of life that is possible only in the country. And even more than that I loved Old Gray, one of Uncle Dirk's draft horses. Big, strong, with beautiful, intelligent, patient eyes—what a marvelous beast he was! In the evening after work I would sit with him in his stable,

feeding him a sugar beet. With my ear against his head, I would listen to the grinding of his teeth, breathe in the strong, horsey smell of his hide, and feel the wonderfully soft lips of this enormous beast,
or, sitting on the driver's seat, decide that his short, twitching tail resembled the head of a long-haired woman shaking her head from an upper-story window.
I'd marvel over what a big "lake" my friend could make
and how much wind he could pass, twenty times, one after the other, one with every step!

And when it was cold outside, Old Gray, puffing away, looked like a locomotive in full steam!

And then, the most glorious times of my boyhood—while riding Old Gray to the blacksmith to have him reshod, to meet someone from my class! What joy I found on the farm!

Nowadays when we occasionally take a trip to Flakkee, I can count all the horses of the entire island on my two hands...

That's why I wrote this book... about everything that is still to be found on a farm before it perhaps changes.

A Brabantine harness
for heavy draft horses

This is how horses
are harnessed in
Zeeland and
Flakkee.

pulley
In Zeeland they use only one rein—one strong pull means left, and a few short jerks mean right. When two horses are in tandem, only the lead—the left horse—is reined; the other one follows automatically.
f you don't have a
ree hand you can
ell the horse where
o go by voice command:
"hiot" for right,
"haar-op" for left.

Gelderland horses

This double-edged plow can be reversed when the farmer reaches the end of the field.

Farm Tools

beetroot cutter
seeder

white onions
small seeder used in planting onions

seedbed frame
MAASLAND
MAASLAND

harvesting
potatoes

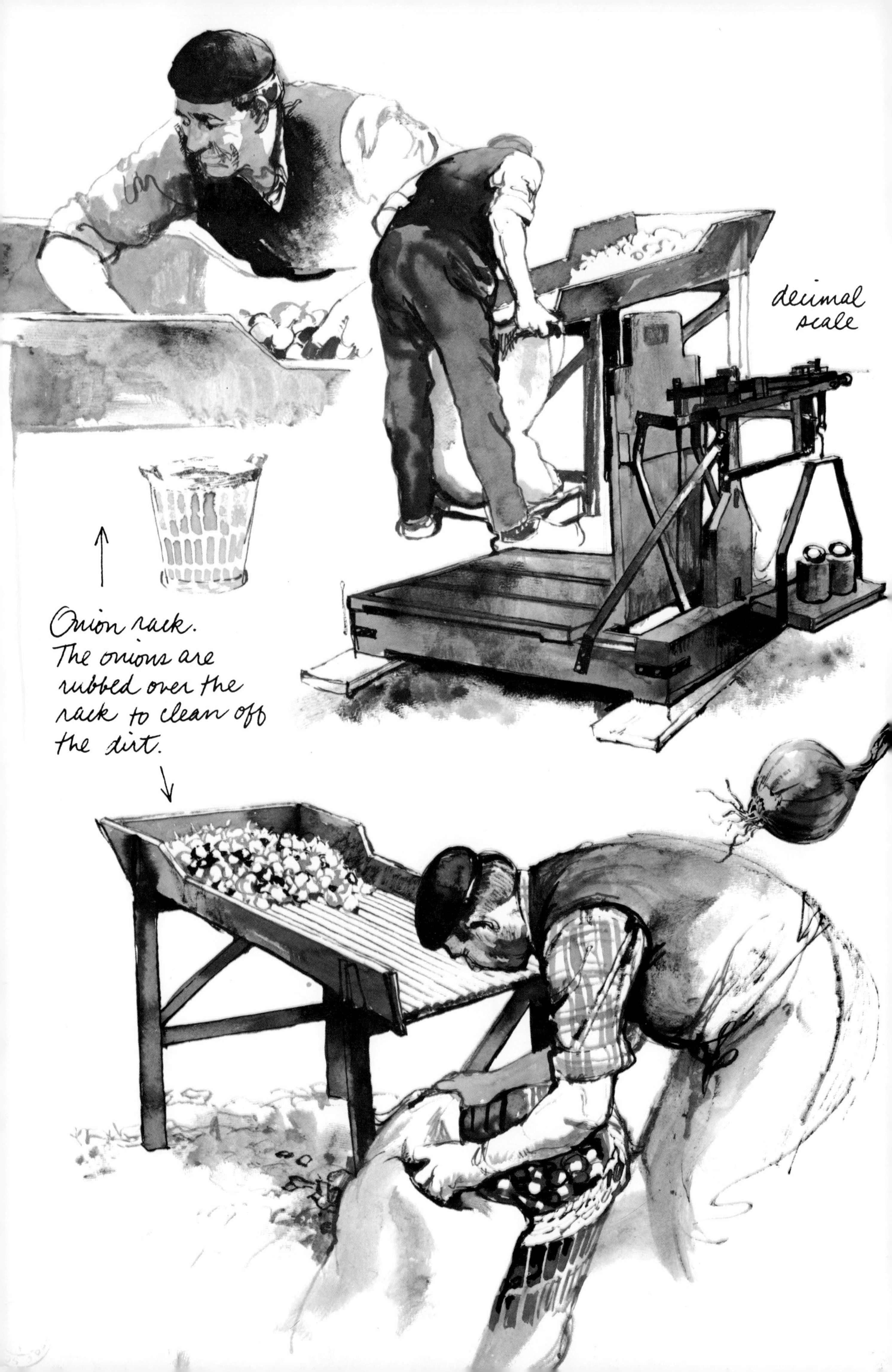
decimal scale
Onion rack. The onions are rubbed over the rack to clean off the dirt.

winnow used to separate the chaff from the wheat
whetstone
village cattle fodder cooperative

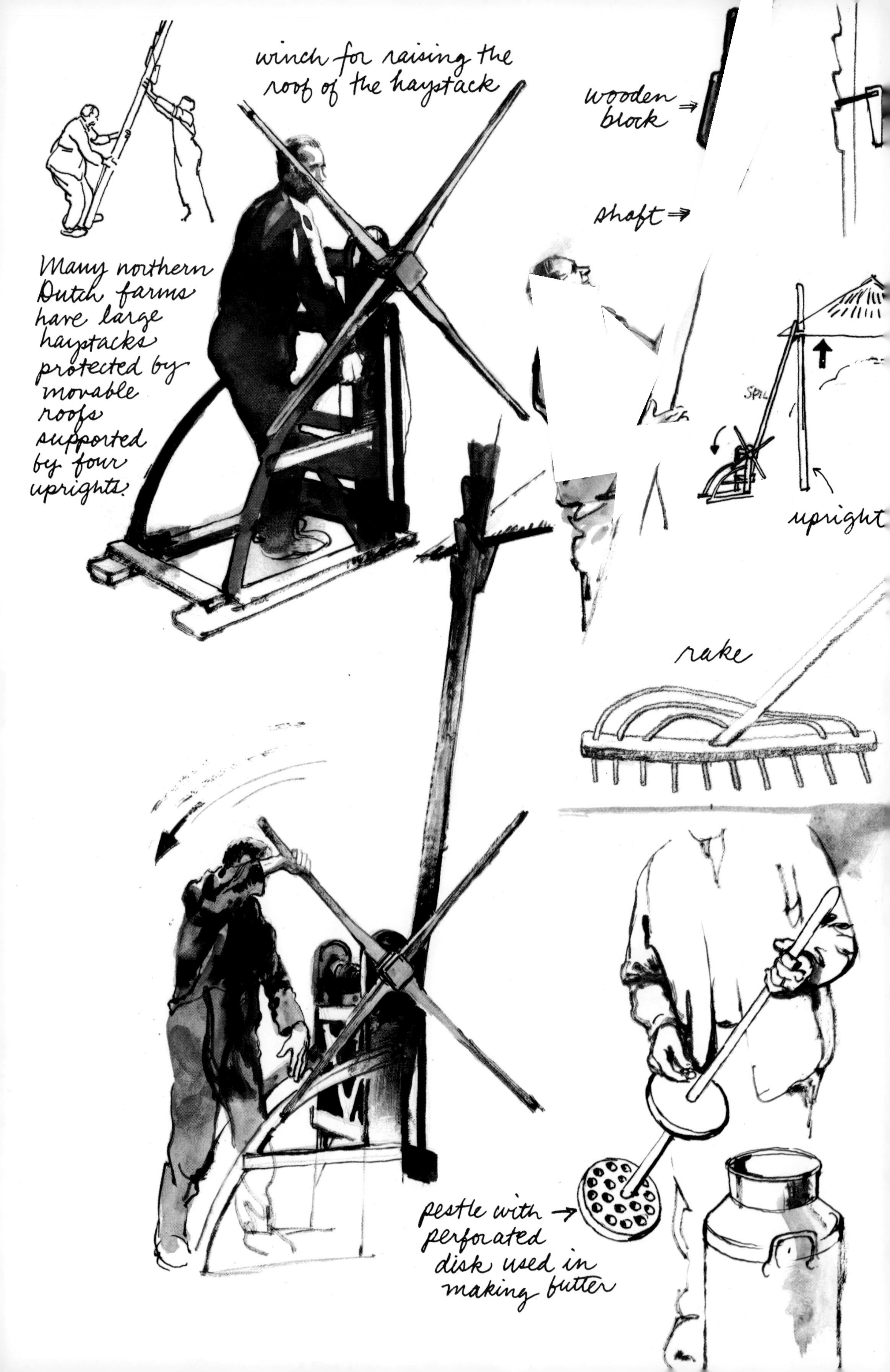
winch for raising the roof of the haystack
Many northern Dutch farms have large haystacks protected by movable roofs supported by four uprights.
wooden block
shaft
upright
rake
pestle with perforated disk used in making butter

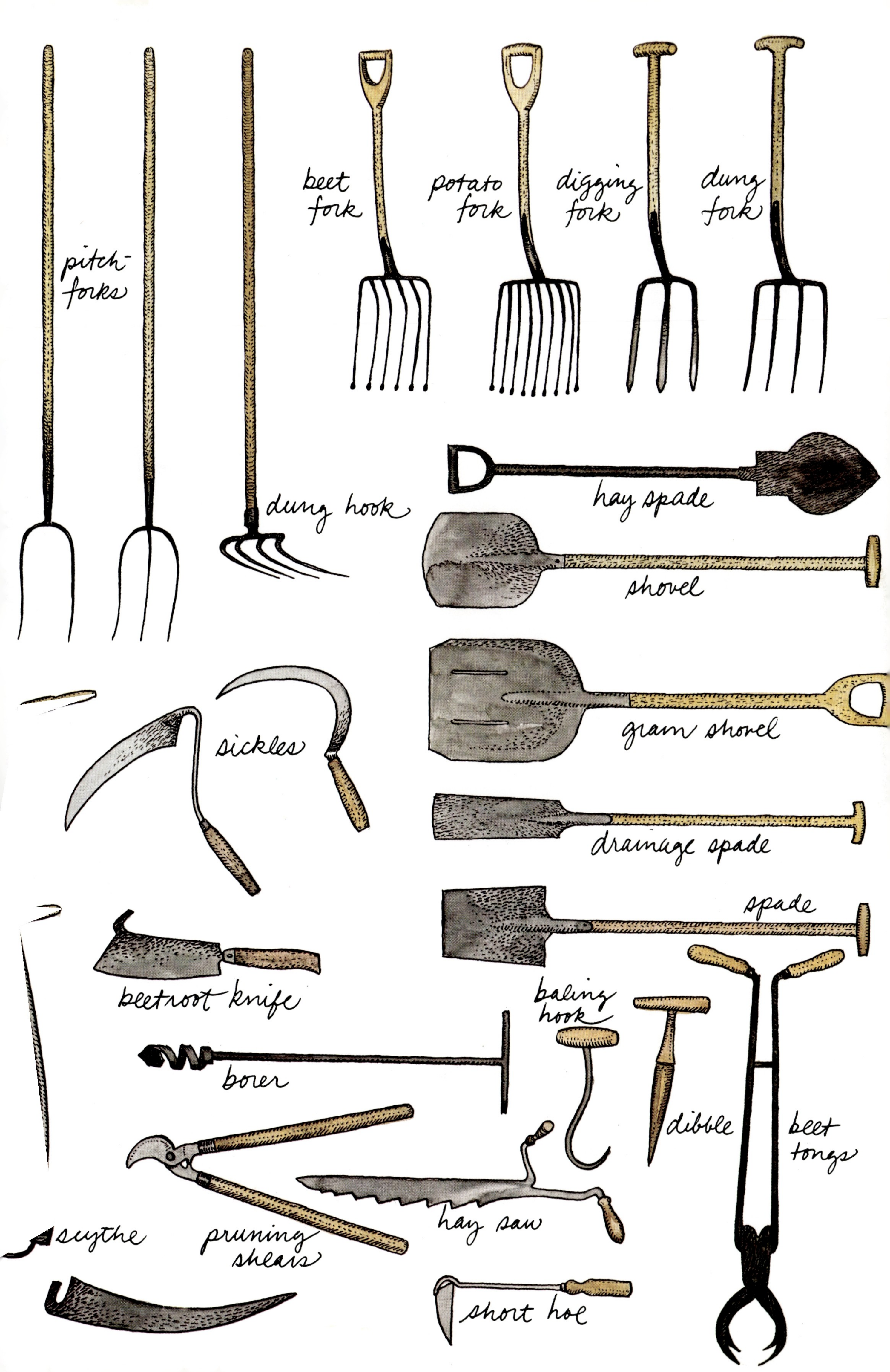
pitch-forks
beet fork
potato fork
digging fork
dung fork
dung hook
hay spade
shovel
sickles
grain shovel
drainage spade
spade
beetroot knife
baling hook
borer
dibble
beet tongs
hay saw
scythe
pruning shears
short hoe

haying

draining
edging
water
and
swill
scoop
sharp metal edge cuts
tree limbs with
the grain of
the wood
It is
better to
prune
trees
than to
saw off
branches,
for rain can easily
penetrate the stumps.
sledgehammer

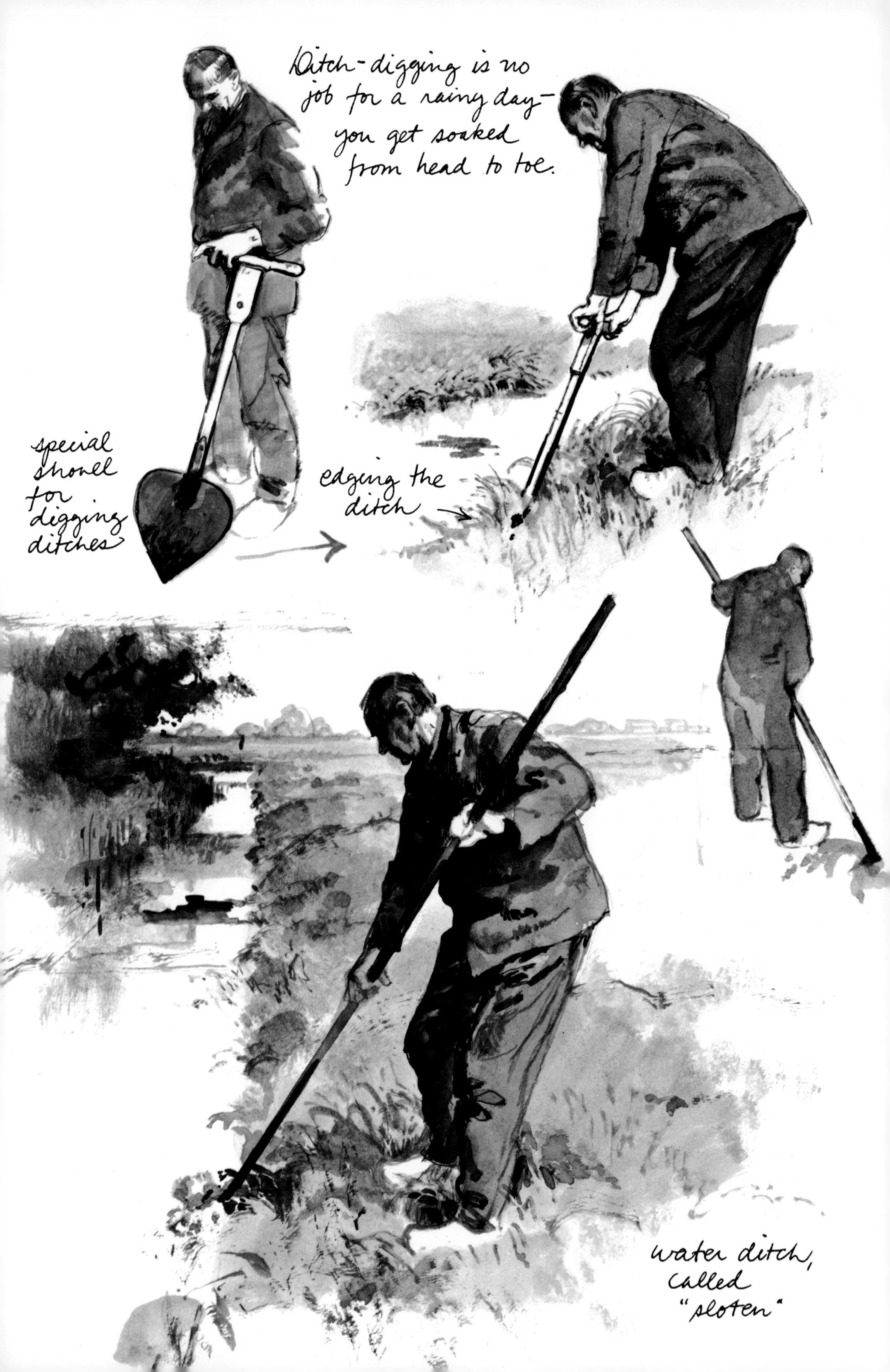
Ditch-digging is no job for a rainy day– you get soaked from head to toe.
special shovel for digging ditches
edging the ditch
water ditch, called "sloten"

hacking the plants along the ditch edge
dredging
A farmers' saying:
Thistle cutting is like thistle growing,
thistle pulling is like thistle breeding,
thistle mowing is like thistle sowing.
wooden tongs for pulling thistles

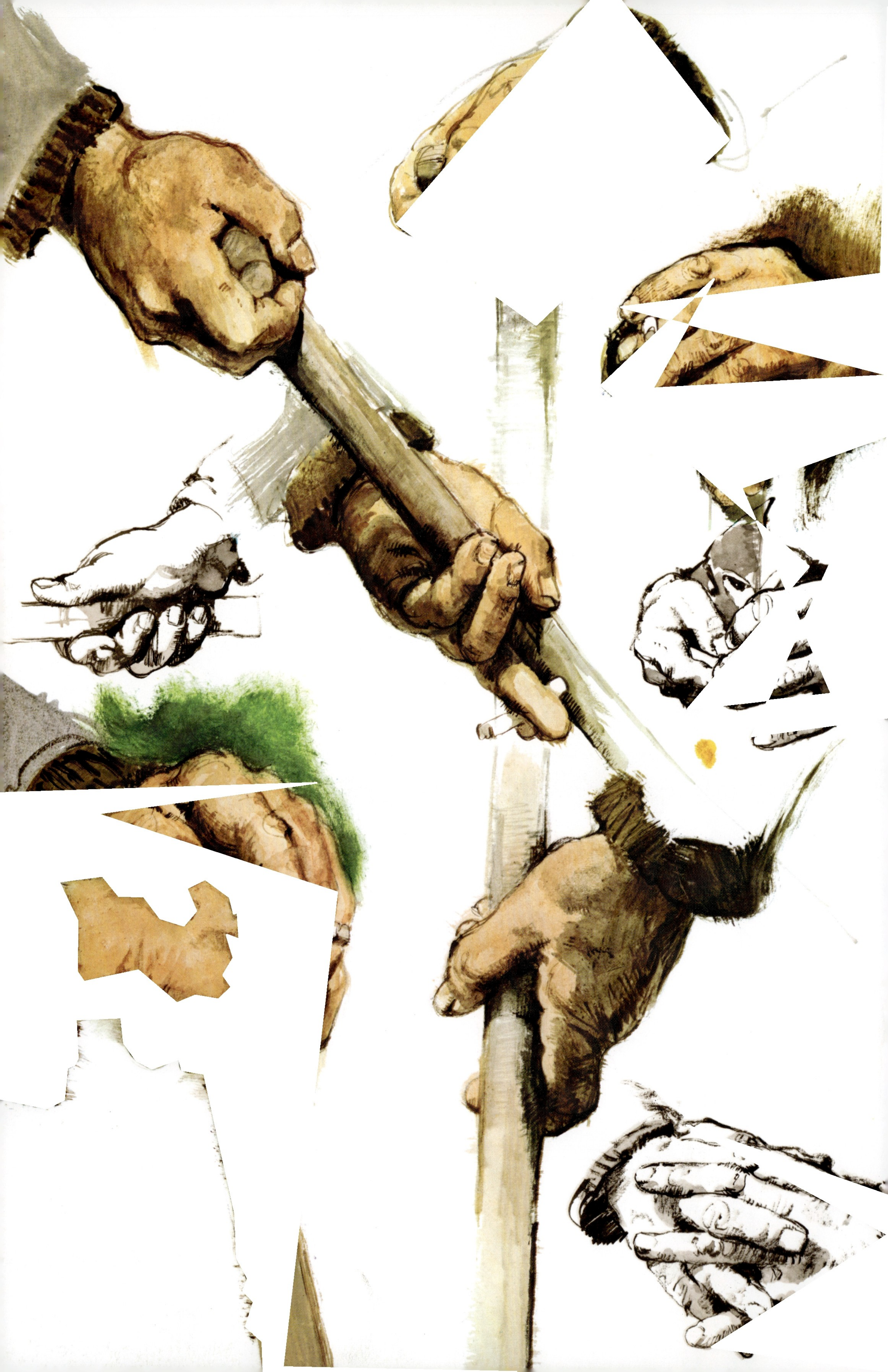

In Drente, mowers and reapers use a "haarspit" (honing anvil) to sharpen their scythes. It is an iron pin with a flat steel head on which they beat the edge of the scythe with another tool called a "haarhamer."

plowing

Turning around
at the edge
of the field

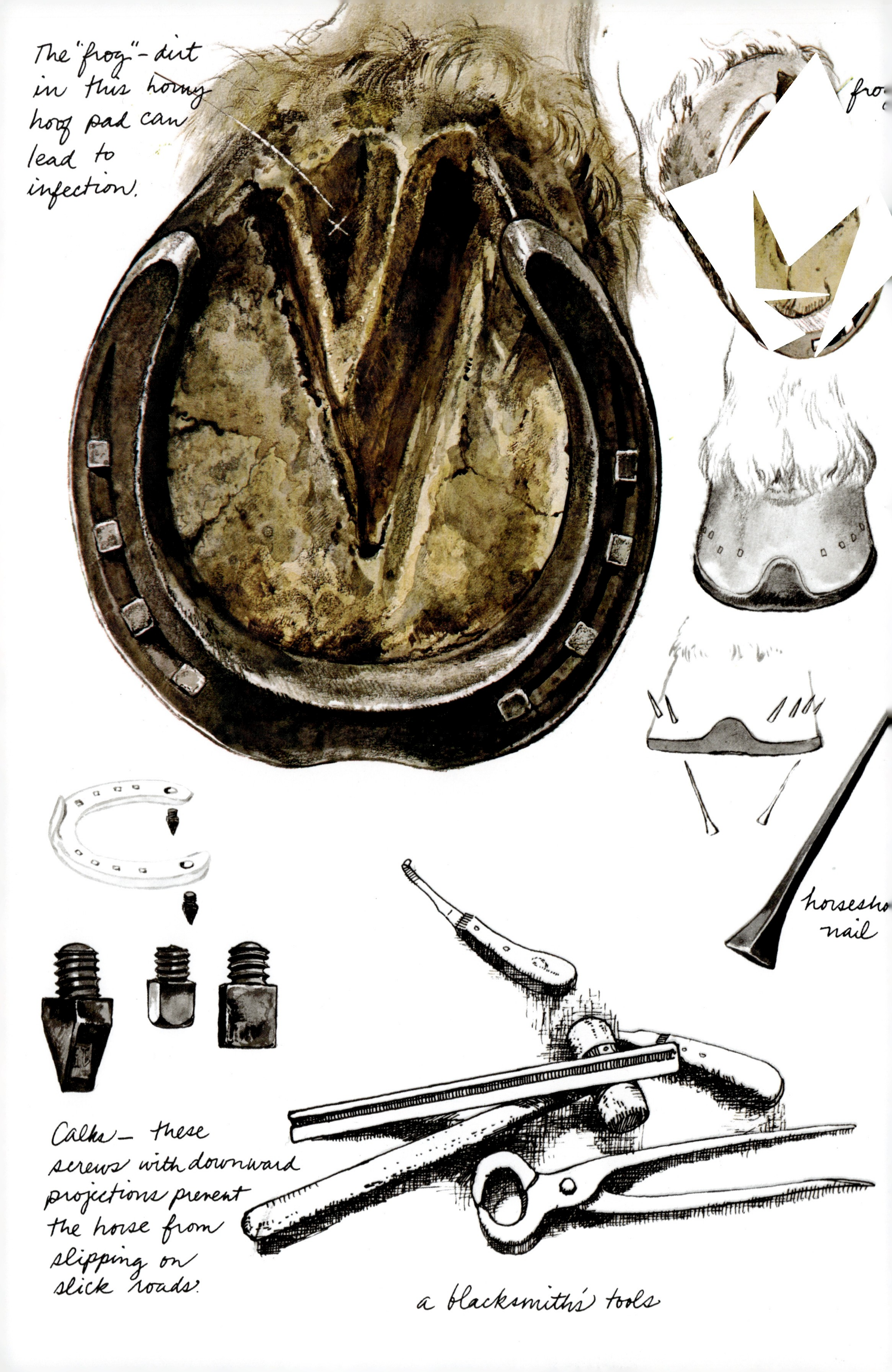
The "frog"- dirt in this horny hoof pad can lead to infection.
frog
horseshoe nail
Calks- these screws with downward projections prevent the horse from slipping on slick roads.
a blacksmith's tools

Horseshoeing

The old horseshoe is removed.

shoeing cage at the smithy

Barn and Barnyard

watchdog
pig pen or sick bay
pump
rack for dripping milk pails
milk cart
pig pen
hearth
bakehouse
kitchen
T.V.
hallway
closet
closet
oven for cooking pig swill
preserves
W.C.
potatoes
maid's chamber

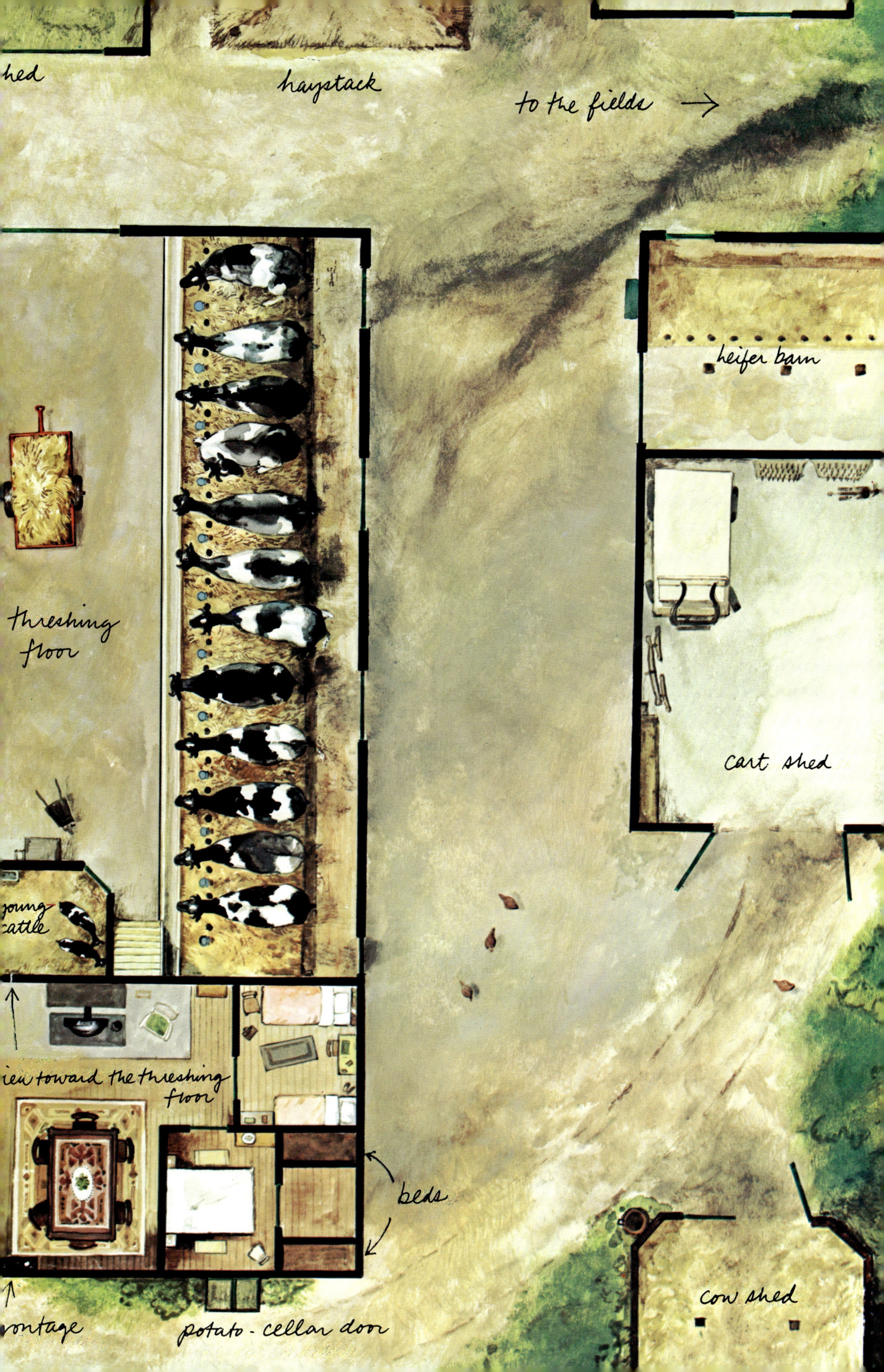

hed
haystack
to the fields →
heifer barn
threshing floor
cart shed
young cattle
iew toward the threshing floor
beds
ontage
potato-cellar door
cow shed

this is the farmhouse shown on the preceding pages

it also is the farmhouse referred to in these old documents.

Transport ƒ	243	50
Twee Stoelkussens op Een gulden Vyftig Cents „	1	50
Een Vyzelen Stamper op Een gulden vijftig Cents „	1	50
Een Kaasvat op vyftig Cents		50
— In de Kelderkamer —		
Een Beddepan op Een Gulden Vyftig Cents	1	50
— In de Geut —		
Een Groote IJzere Pot en deksel en Vier kleinere dito op Tien Gulden „	10	„
Twee tinne Schotels op Vyf Gulden „	5	„
Een Groote Kopere Schenkketel en Vier kleinere dito op Tien Gulden „	10	„
Een Melkkarn en tree op Zes gulden „	6	„
Een Vatebank met eenige aarde Potten en Pannen op Drie gulden „	3	„
Twee Provisiekastjes op Drie Gulden „	3	„
Een Eijerrek en Zes Kruiken op Dertig Cents „	„	30
— In de Kelder —		
Zesendertig roompotten op Tien gulden „	10	„
— In het Karnhuis —		
Een Melkkarn tree en Twee melkstoppen op Zes gulden „	6	„
Eenige tonnen en tobben op Drie Gulden „	3	„
ƒ	304	80

On June 13, 1826, the notary public Frans Pen of Baarn drew up a document on behalf of the heirs of Arnd Sansen van Klooster and his wife, Sannetje Anthonuisje van Logtestein. It contained an inventory and description of the farm equipment, livestock, jewelry, furnishings, etc., in the house and on their land, an estate called The Cloister—the very same one as on the facing page. Above is a page of that document.

cellar door ↑ cellar stairs ↑

Leghorn fowl

the old bakehouse in which bread was baked

These are special Dutch breeds of poultry. They are not particularly important commercially, although the leghorn, originally a hardy fowl from Italy, is an excellent egg layer.
speckled leghorn
Dutch gray
leghorn "sorting chick:" this means that as soon as the chick is hatched it is possible to tell whether it's a rooster or a hen.
white leghorn
Dutch russet
North Holland blue

Barneveld fowl
russet leghorn
Dutch rabbit

Chicken
coops

With this type of henhouse the chickens can sleep indoors!

Dirksland

The geese guard the farmstead even better than the watchdog.

piebald pig

Frisia
Eieren
18 cent

oats
rye
wheat
barley

Some farmers don't feed their cats; this makes them better rodent hunters but then they also start poaching.

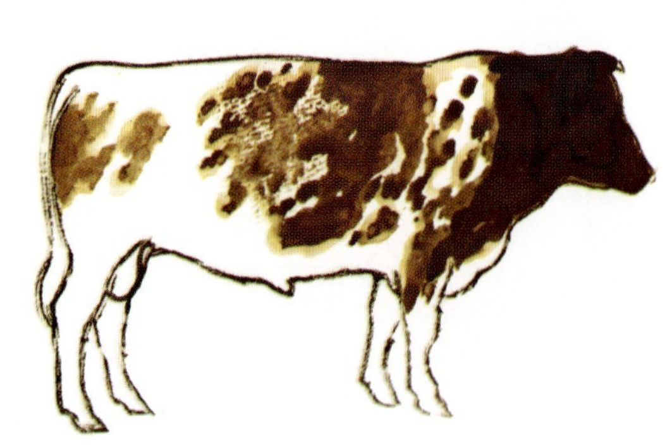

Black-and brown-spotted breeding cattle

Weights range from 2,000 to 2,200 lbs.

Groningen cattle, usually beef rather than dairy

Lakenvelder cattle – a breed, rarely seen now, that used to be kept in the meadows of estates. "Laken" refers to the "sheet" that looks as if it were stretched across its back.

ear tag
branding tool with sharp notches for identifying numbers and letters

droop-eared Landrace pig
Yorkshire boar

a pretty scrawny cow
some horses have the nasty habit of gnawing on their trough or fence.

cow barn in a Frisian farmhouse

Tether—in an emergency, such as a fire, it can be opened in seconds.

casionally a cow's horns will
ow into the animal's head, in
hich case a piece is
awed off.
Tying a
cow's legs...
nd
hen
ilking
er,
queezing
it the last
drop.

Before milking, the udder
must be cleaned. Two
daily milkings yield
approximately 20 quarts.

Annie 7
milking stool
milk
strainer

milk pails
weights
cheese press
perforated strip
(or cheesecloth)
used to separate
the curds and
whey
skimming
off the
cream
cream tub
cheese vat
kneading the
curds in the
cheese vat

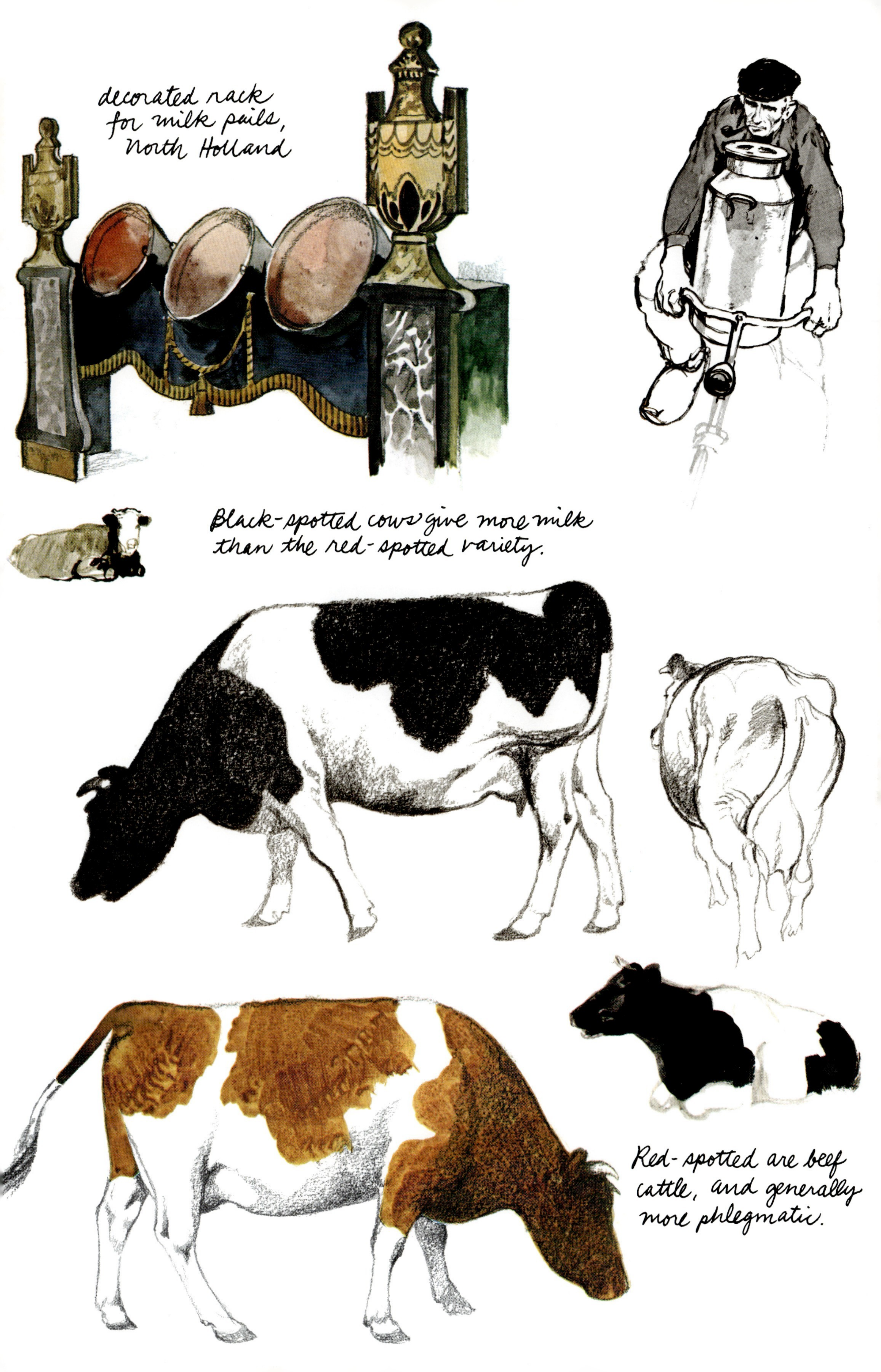
decorated rack
for milk pails,
North Holland
Black-spotted cows give more milk
than the red-spotted variety.
Red-spotted are beef
cattle, and generally
more phlegmatic.

Chewing the cud means that the grass is ruminated about fifty times.
the semi-annual pedicure
paring knife

taking measurements
6,014 for a two-year-old cow
6,250 thereafter
fat content
number of days it took for the 6,014 and 6,250 yield
albumin content
points for appearance
4.28 - 305 3.37
4.26 - 336 3.49
86 pnt
Gretha 87
6014 4.28 - 305 3.37
6250 - 4.26 - 336 3.49 86 pnt
ribbons for prize-winning cattle

a cow with the habit of stepping on her udders when getting up

a cow with the habit of sucking on her neighbor's udder. This studded band puts a stop to that.

What to do if your cow falls into a ditch or canal: if you try to pull a cow by a rope tied to its horns it will resist stubbornly, but if you tie the rope around the cow's neck and pull, it can't get air, and although resisting, can be pulled out.

But, if at all possible, the best way is to pull her out legs first.

A "ditch hurtler."
But with this small log
across her knees she won't
do it again.

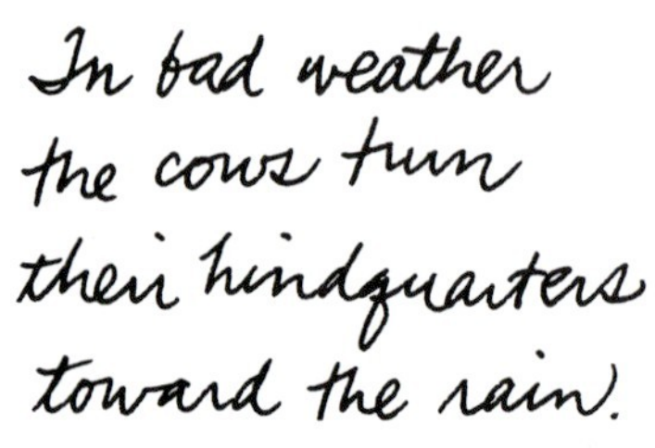

In bad weather
the cows turn
their hindquarters
toward the rain.

Here the manure is pushed through the grate.
In winter cattle are stabled and fed twice daily with hay, enriched feed, grain, or silage.

haying machine

81

Mating and Birth

A stamp pad tied
around the ram
shows at a glance
which ewes have
been serviced.

Sustained good, dry weather makes the mares more receptive and their pregnancies easier.

Cows come into heat every three weeks. The bulls then mount them.

Mating takes
two seconds—
up and down
A cow's gestation period
is about nine months.

By touching each side of the root of the tail, one can feel the "bands" loosen, a sign that labor is about to begin.
Before it starts, the cow lies down.
First comes the amniotic sac, and then the legs.

Sometimes the calf has to be pulled out.

The calf is rubbed dry with handfuls of straw.

taking the temperature and listening to chest noises
the veterinarian
To distract the animal the vet grabs it by its nostrils before giving it a shot.
rubber suction hose in a pail of mother's milk

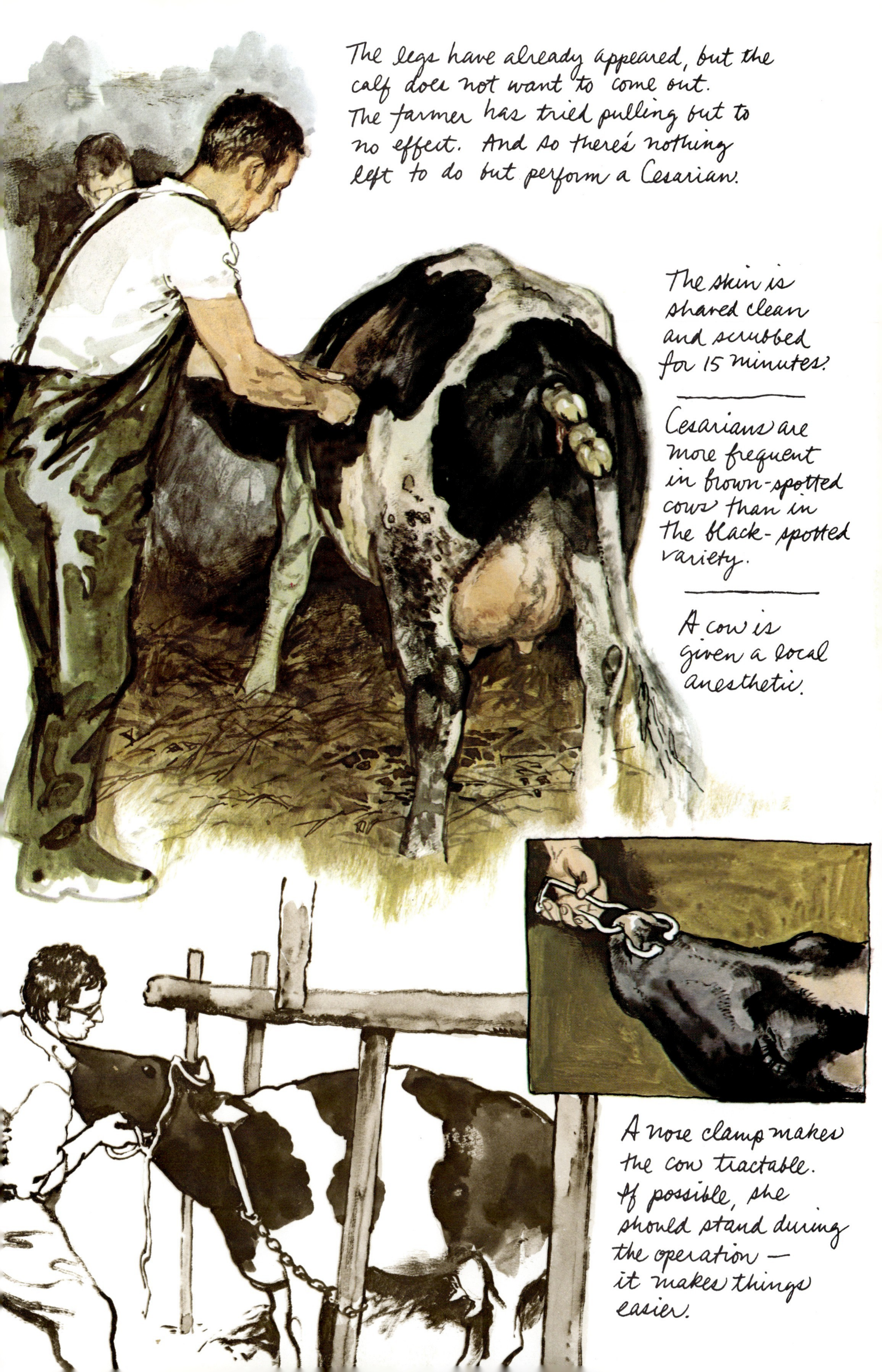
The legs have already appeared, but the calf does not want to come out. The farmer has tried pulling out to no effect. And so there's nothing left to do but perform a Cesarian.
The skin is shaved clean and scrubbed for 15 minutes.
Cesarians are more frequent in brown-spotted cows than in the black-spotted variety.
A cow is given a local anesthetic.
A nose clamp makes the cow tractable. If possible, she should stand during the operation — it makes things easier.

A ten-inch incision through
the skin, the muscles,
and the peritoneum.
The cow decides
to lie down.
Nothing to be
done about it...
The intestines keep poppin
out throug
the incisio
and this
must be
prevented.
The doctor
gropes for
the uteru

Once the uterus has been laid bare an incision is made in it near the forelegs of the unborn calf.

The calf is pulled out, hale and hearty.

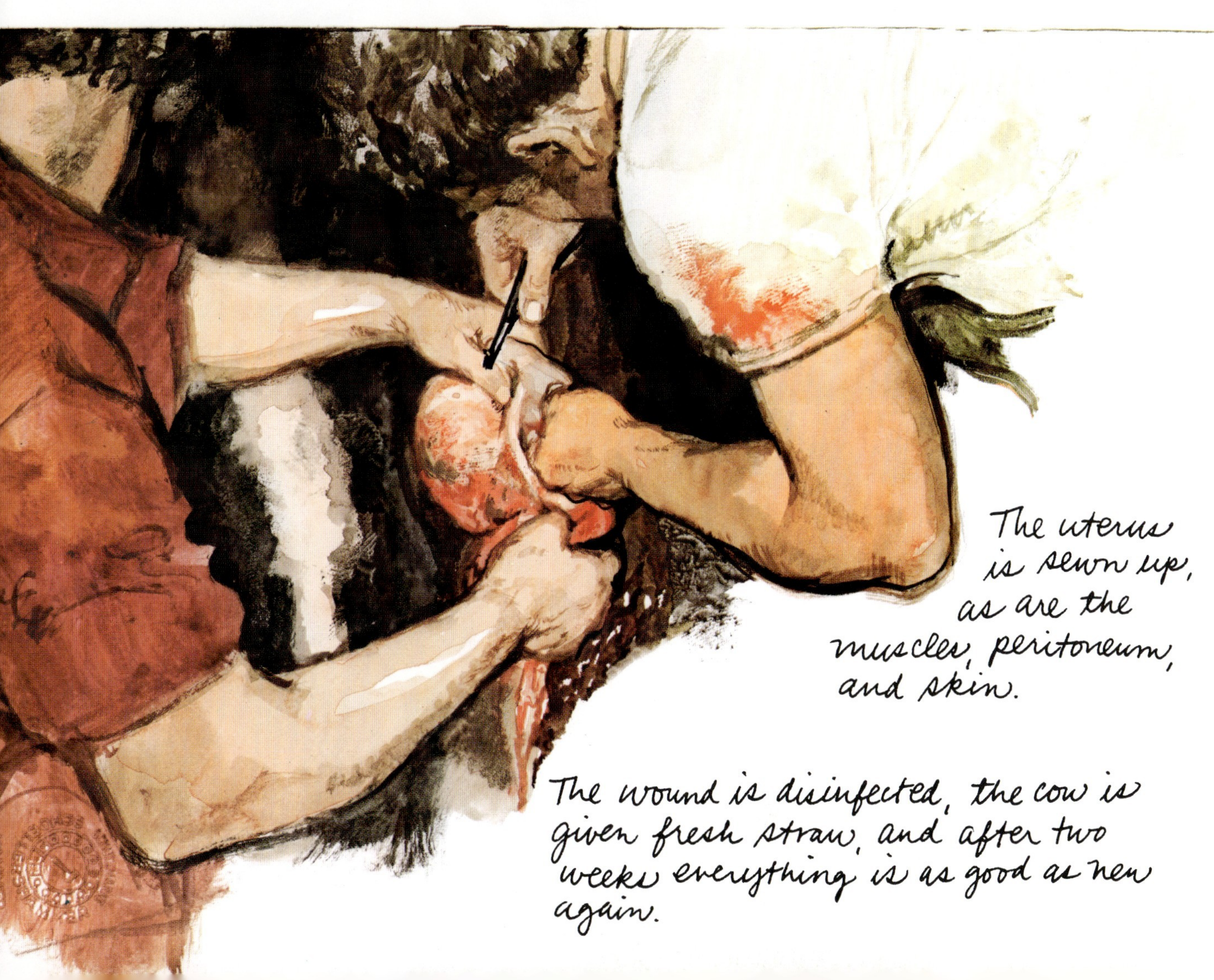

The uterus is sewn up, as are the muscles, peritoneum, and skin.

The wound is disinfected, the cow is given fresh straw, and after two weeks everything is as good as new again.

A day or two before farrowing, the sow is put into the farrowing bay to prevent her from suffocating the piglets.
waiting for labor to begin

She farrows twice a year.
Heating lamp to keep the little ones warm
At the age of eight a pig is already old.
six-week-old piglets

This piglet died shortly after birth. After all, it was one of 15, and the sow had only 12 nipples. A litter that big is bound to have one, two, or three that will not survive. A sow has anywhere from 10 to 14 nipples. According to farmer Lam, by the third day, each piglet has "its own nipple."
The newborn piglet has sharp teeth and the farmer clips them off, to makes things more pleasant for the sow.

The Vet

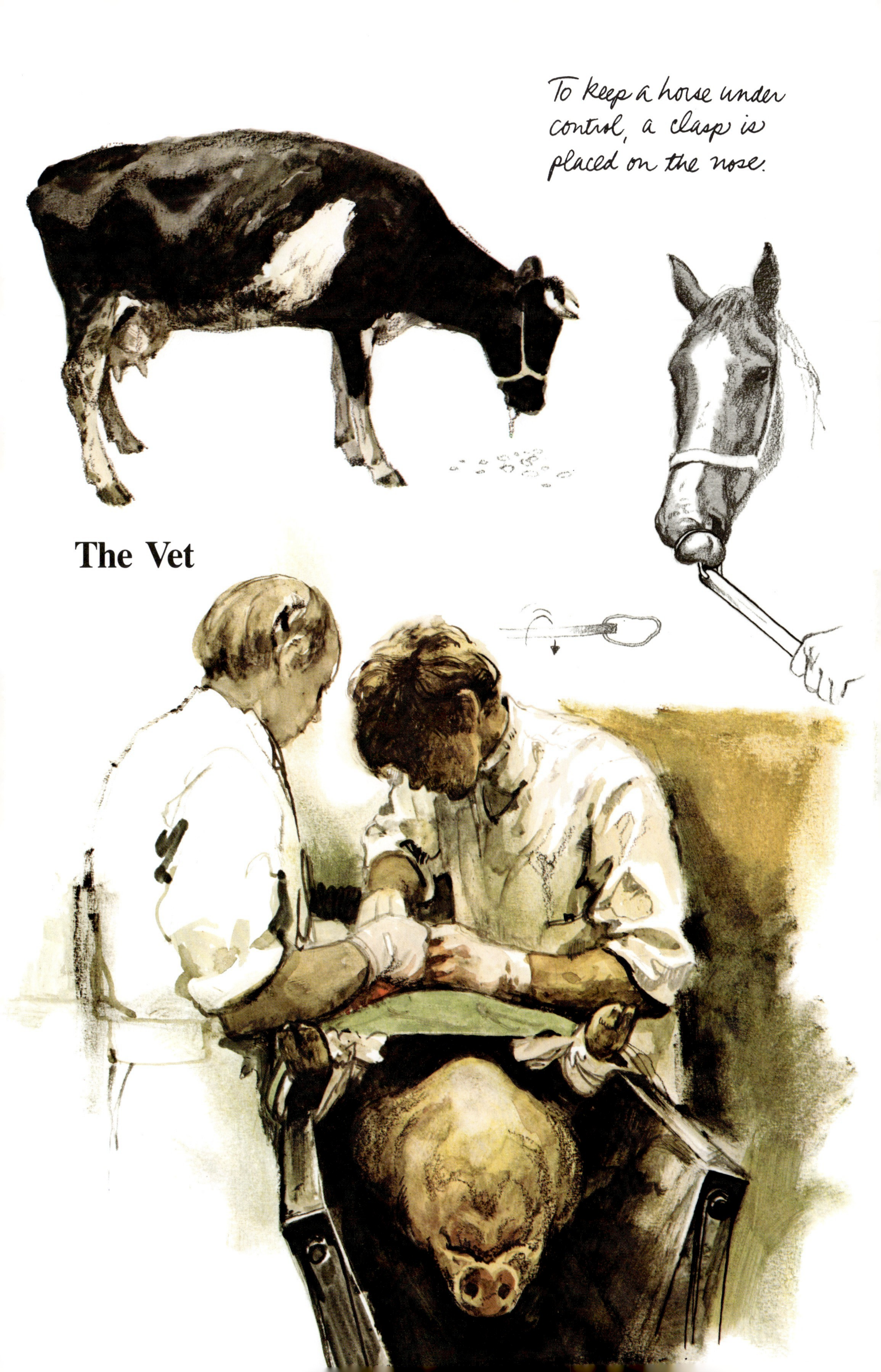

To keep a horse under control, a clasp is placed on the nose.

That's how you tell
the age of a horse.

leg surgery
for a heifer

"Uithof" the school for veterinary medicine of Utrecht
administering the anesthesia

jaw clamp
first, scrubbing up for ten minutes

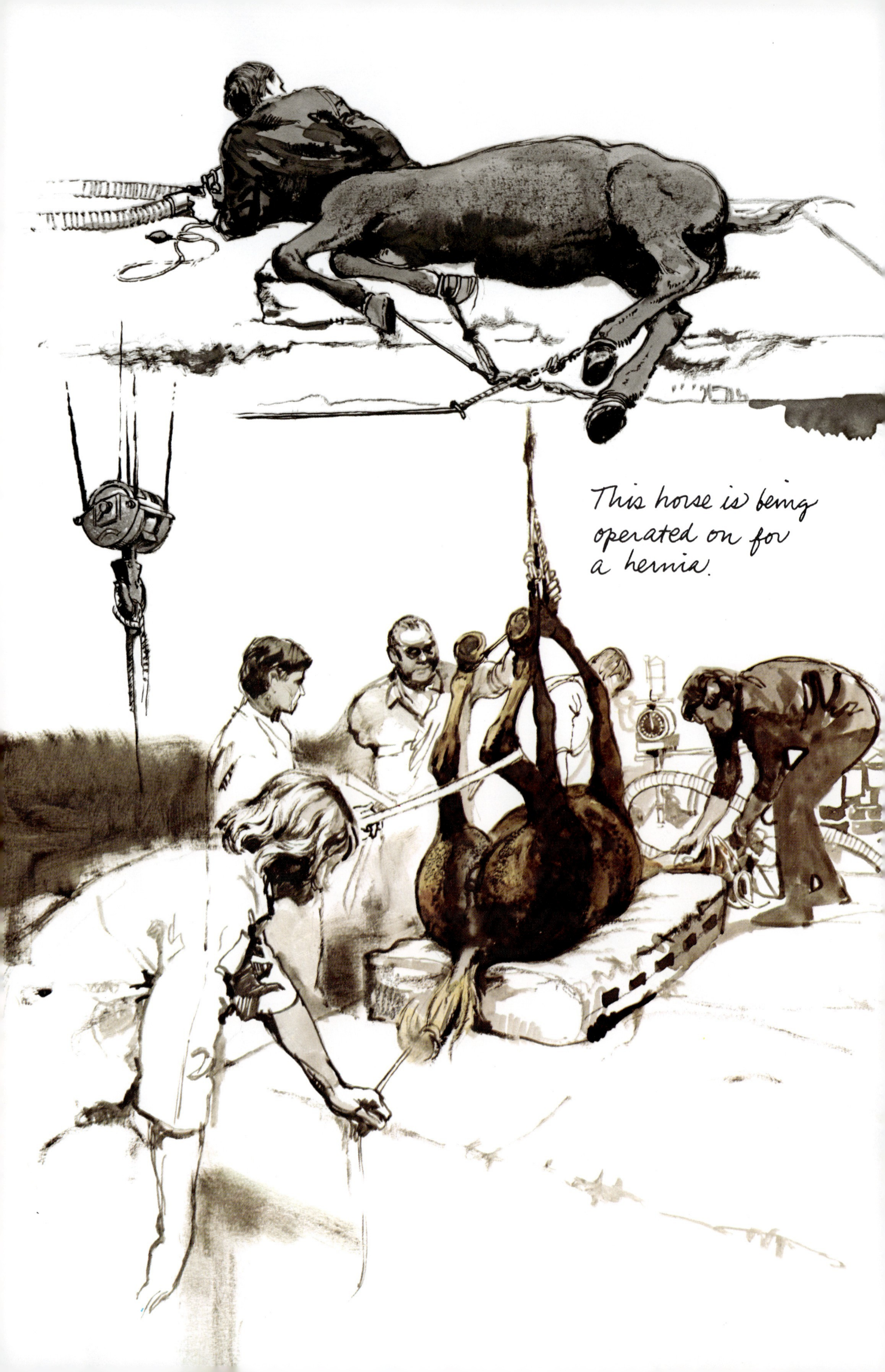
This horse is being
operated on for
a hernia.

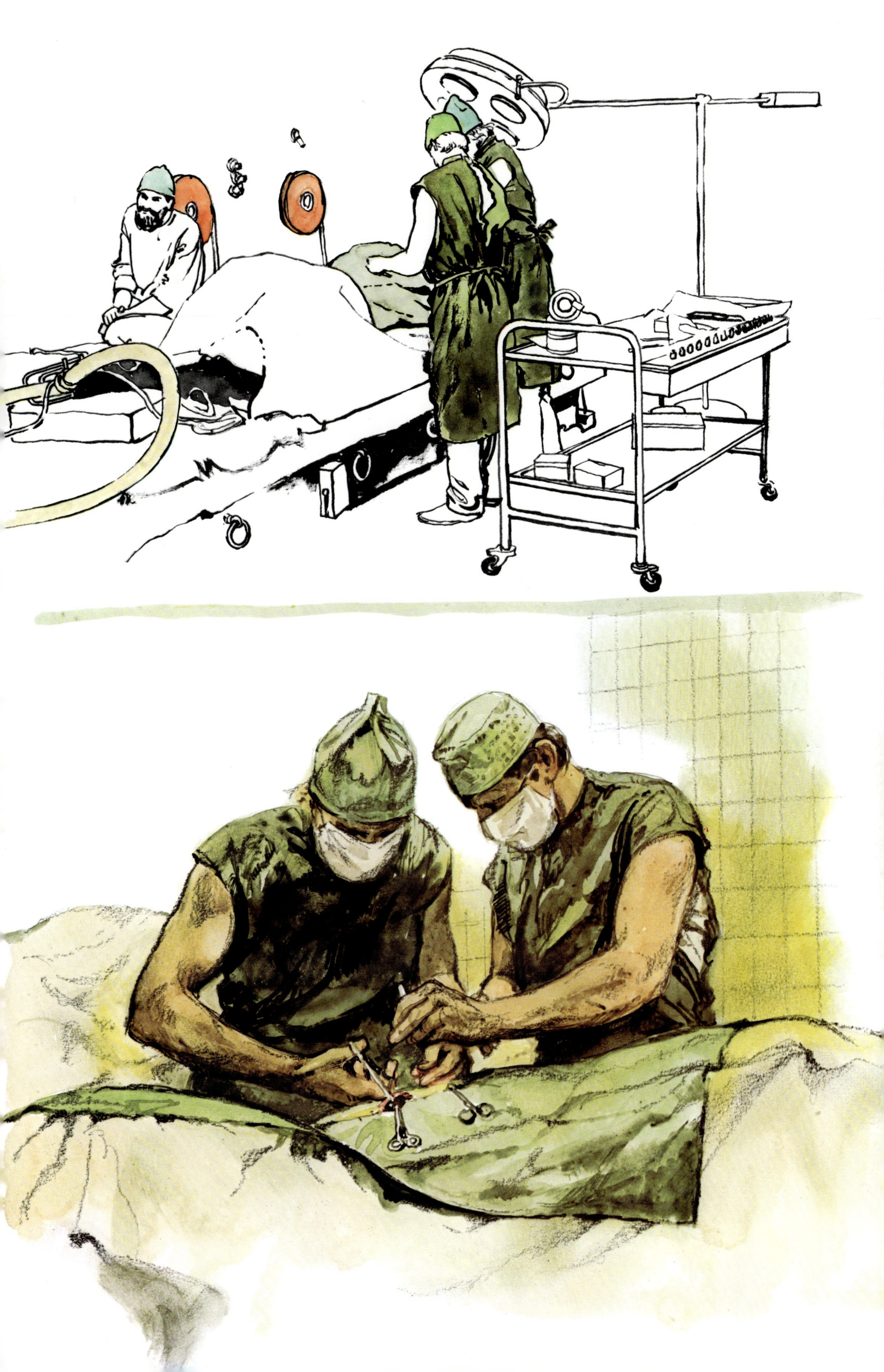

Occupations

the pastor
the teacher

the mailman

the barber

The Clogmaker

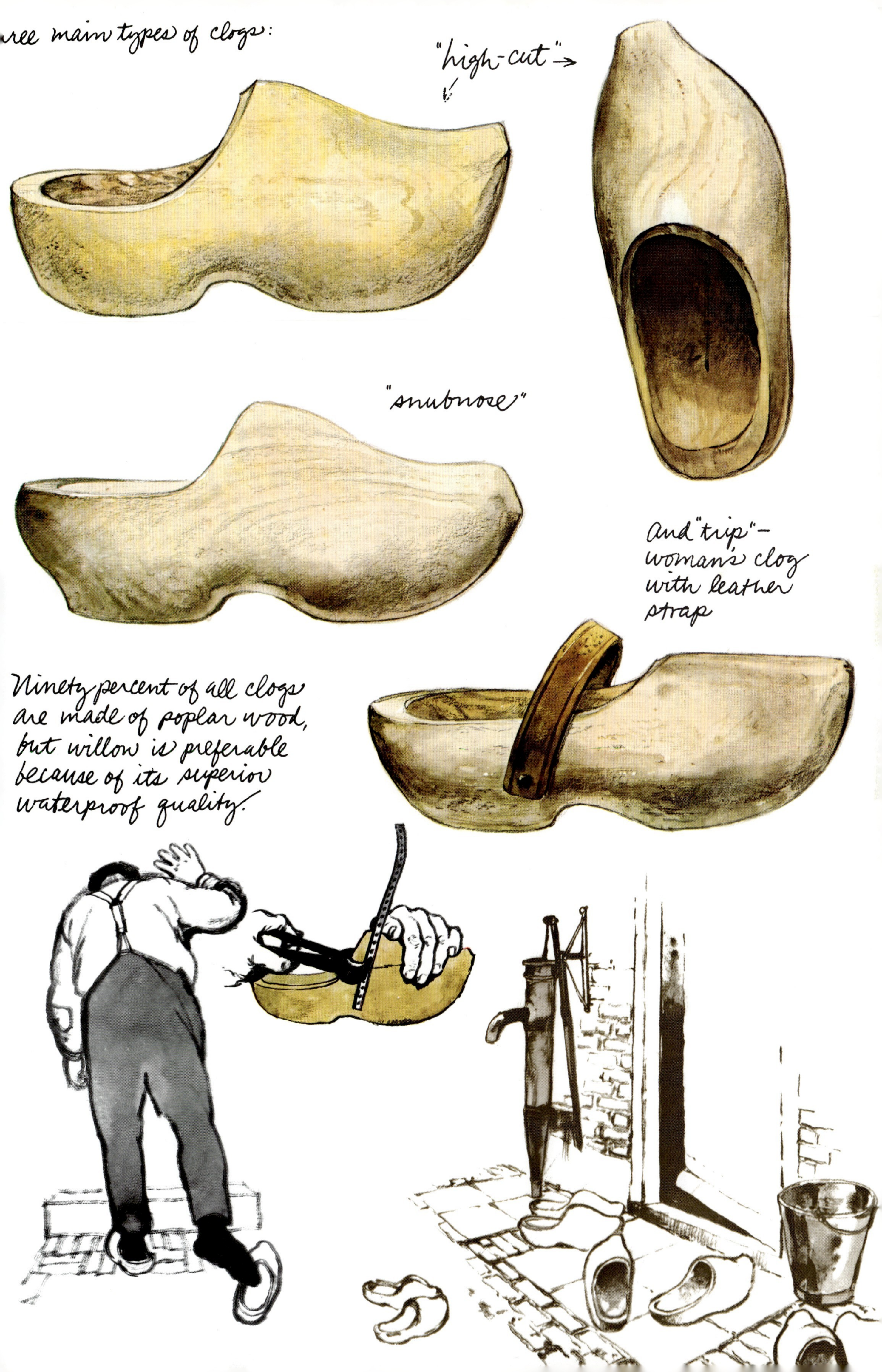
ree main types of clogs:
"high-cut"
"snubnose"
and "tip"–
woman's clog
with leather
strap
Ninety percent of all clogs
are made of poplar wood,
but willow is preferable
because of its superior
waterproof quality.

a good remedy
or cold feet:
Lining the clogs with straw.
cutting a log into 16-inch lengths
Each piece is divided into six sections,
and each section makes one wooden shoe.

Queen's day

typical Utrecht farmhouse

Sirjans land
Chris

A Day on the Farm

Of course no storebought cream for this coffee! It comes directly from the cow, and is then whipped up.

preserving jar used for putting up endives

Zeeland farmhouse

after the meal
a passage from the Bible

a quick nap after lunch
barn owl

a little more enriched feed
mole trap
setting the trap in the mole passageway

something for the rabbits
my great-grandfather Sakries Poortvliet and sons

Some farmhouses with thatched roofs have a tiled inset to catch the rainwater.
A good thatched roof lasts 50-60 years, depending partly on its slope.

in the haystack

gathering eggs

milk cart

The milk is picked up twice daily, but in winter and on Sundays only once a day.

The outhouse — a bone-chilling obligation

wiping off the clothes line

waterwheel regulating the level of the dikewater

Groningen

If by day's end
the work has not
been finished,
the farmer leaves
his plow and harrow
in the field and goes
home on horseback.

If you're had enough coffee you turn your cup upside-down.
listening to the "Agricultural and Gardening News Bulletin"

pocketwatch case

a rack
for drying
long knitted
stockings
Before retiring at night,
the shutters are closed
and latched.

At dusk the bats appear.

Hunting

hunting along the river ysel

behind the house

Keeping animals and birds off cultivated areas is quite a problem.
roe deer
wild duck
hare
partridge
pheasant
rabbit
wild boar
all of them are forever foraging
wood pigeon
wild goose

the kestrel,
an excellent mouser

pigeon hunting
perch in the orchard for shooing away
the birds

In the winter
the wild geese
rrive from the far North.
a cure for
cold hands
That's how you blow
your nose outdoors.

Buying something at the animal fair is a serious matter requiring careful consideration.

Buying and Selling

Before a cow is put up for sale,
the hindquarters are hosed down.

A touch of make-up improves the appearance of the udder.
beauty treatment
branding

bargaining

marking the
animals that
have been sold

When a steer gets to be around 1½ to 2 years, a ring is put through his nose when he is taken to market.

tell the age of
cow, you count
he rings on the
orns and add two—
ecause a cow calves
r the first time
age two, after
hich come the rings—
but with
steer, look
at the teeth.

Farmhouses

the brake

Goldereede

farmhouse on the
Zievent-Ruurlo road,
nicknamed "119th Psalm"
because of its great length

St. Nicholas and his helper Black Pete

sloped-roof farmhouse in Hitzum, Frisia

farmhouse in Ervijk with centered threshing floor in rear

Long, gabled house in Waalre, North Brabant.

Zeeland farmstead

Maasland

farmhouse in Wiering, North Holland

Sinaeda Estate
farmhouse in Tzummarum, Frisia

latched
shutters

Groningen
Frisia

Langbroek
farms in the eastern part of Holland
rear view
sheep barn

Flemish barn group and threshing chamber, South Beijerland

house with threshing chamber in the middle, South Beijerland

Chaam,
North Brabant
I so much wanted to draw
the interior, but I wasn't
allowed inside. Fortunately
some lambs were standing
about outside—

they
did not
object.

Maas-land

In the outskirts of Einschede

ANWB

The Village

Heineken
bier
CAFE
A. DE WEERT
Staphorst
mourning
dress is
worn
for four years after the dea
of a parent; for two yea
after the death o
a brother, sister,
and grandparent
for four months
after the death of an uncle
and aunt; for eight weeks
after the death of a nephew
niece, and acquaintance.

福山酒樓
VOLLEDIGE VERGUNNING
"Not on your life!" Dutch farmers usually don't like to try unfamiliar cuisines. Their saying is: "What I don't eat can't make me sick."
above the door, a traditional "force of life"

Kethel

Ouddorp, South Holland

Soest

Frisia

North Holland

Tzummarum

Dutch Reformed Church, Kethel

This is how to keep your coat from getting wrinkled.

KING
the collection

a wedding

Saturday outfit–
the suit jacket
from your best
suit

Saturday evening:
promenading on Main Street

SAMSON
HALFZWARE SHAG

My brother
Harm,
and this is where he lives.

A Woman of Bunschoten

bodice
The pocket is tied around the waist with a ribbon,
and over it goes the black skirt with a slit for the pocket.
stiffly starched bib
Rolled cuffs are pinned to the sleeves of the bodice. All together, eighteen pins are needed for the outfit.

This woman has about 20 bibs, each with a different design. They are hand painted in Bunschoten.
The bib is fastened front and back with ties.
Then a red cloth is pinned to the front and back.

The little cap is crocheted and starched.

a closing word:
Without the friendly help of all the men and women I met on my wanderings about the countryside I certainly could not have done this book.

On many of the farms I was allowed to poke around to my heart's content.
Some people were even nice enough to call me up to tell of special things to see, such as someone plowing a field with four horses. →

The phone rang night and day: "The sow is about to farrow."

One Sunday morning just as I was leaving for church with my family, the phone rang: "The bull is ready to mate if you want to see it."
"Yes, but I'm just leaving for church." So the farmer said, "Well, we'll wait for you."

That was very nice of the farmer, don't you think? Not to mention the bull.

In the face of so much friendliness one can even forget the nasty watchdog that chased me...

or the goat that merrily ate my whole day's work,

↑

and I gladly settle for the large cups of well-meant coffee, sometimes with milk curds as big as undershorts. That's part of it. Joking aside, without all this help I could never have done it. To all of them my warmest thanks!

To my delight, I was given an old-fashioned farmer's suit during one of my visits. It's a shame the moths had already done their work, but I only noticed that when I got it back from the cleaner's.

Well, don't look at it too closely. I might even wear it to church one day.

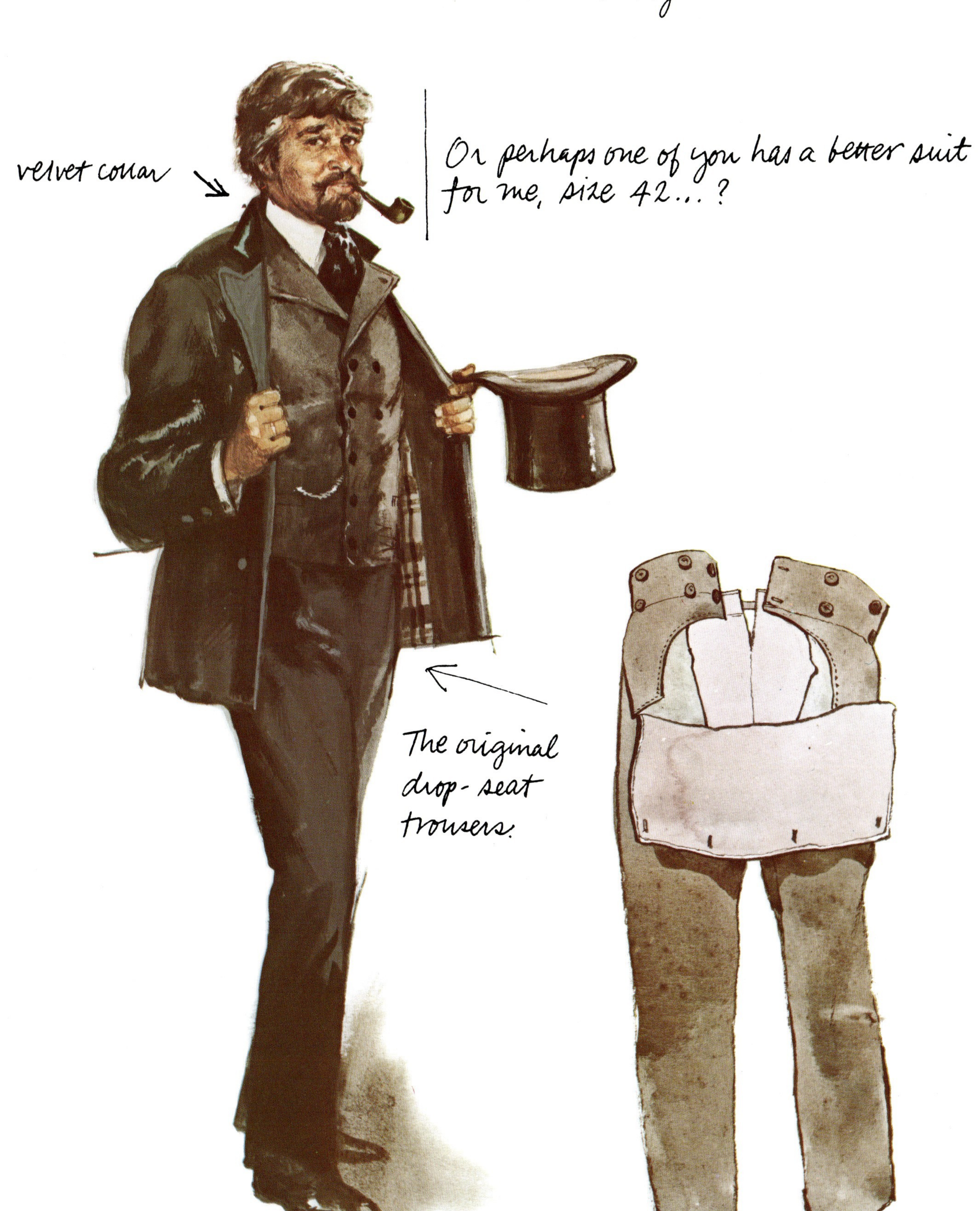

Or perhaps one of you has a better suit for me, size 42...?
velvet collar
The original drop-seat trousers.

watchdog
pig pen
or
sick bay
pump
rack
for dripping
milk pails
milk cart
pig pen
hearth
bakehouse
kitchen
T.V.
hallway
closet
closet
oven
for
cooking
pig
swill
preserves
W.C.
potatoes
maid's
chamber